Contents

	PAGE
Section 1: Introduction	2
What is an Approved Document?	2
Consideration of technical risk	2
How to use this Approved Document	2
Where you can get further help	3
Responsibility for compliance	3
Section 2: The Requirements	4
LIMITATION ON REQUIREMENTS	6
Section 3: General guidance	7
Key terms	7
Types of work covered by this Approved Document	8
Buildings exempt from the energy efficiency requirements	9
Notification of work covered by the energy efficiency requirements	11
Materials and workmanship	12
The Workplace (Health, Safety and Welfare) Regulations 1992	13
Section 4: Guidance relating to building work	14
EXTENSIONS	14
Large extensions	14
Other extensions – reference method	14
Optional approaches with more design flexibility	14
Conservatories and porches	15
Swimming pool basins	15
MATERIAL CHANGE OF USE AND CHANGE OF ENERGY STATUS	15
Material change of use	15
Change of energy status	15
WORK ON CONTROLLED FITTINGS AND SERVICES	16
Controlled fittings	16
Controlled services	17
COMMISSIONING OF FIXED BUILDING SERVICES	19
Notice of completion of commissioning	19
Section 5: Guidance	
THE PROVISION OF THERMAL ELEMENTS	21
Continuity of insulation and airtightness	21
RENOVATION OF THERMAL ELEMENTS	21
RETAINED THERMAL ELEMENTS	22
Section 6: Consequential improvements	24
Consequential improvements on extending a building	24
Consequential improvements on installing building services	25
Section 7: Providing information	26
Building log book	26
Appendix A: Documents referred to	27
Appendix B: Standards referred to	28
Index	29

Approved Document L2B

Conservation of fuel and power

Section 1: Introduction

What is an Approved Document?

1.1 This Approved Document, which takes effect on 1 October 2010, has been approved and issued by the Secretary of State to provide practical guidance on ways of complying with the **energy efficiency requirements** (see Section 2) and regulation 7 of the Building Regulations 2000 (SI 2000/2531) for England and Wales, as amended. Regulation 2(1) of the Building Regulations defines the **energy efficiency requirements** as the requirements of regulations 4A, 17C, 17D and 17E and Part L of Schedule 1. The Building Regulations 2000 are referred to throughout the remainder of this Document as 'the Building Regulations'.

1.2 The intention of issuing Approved Documents is to provide guidance about compliance with specific aspects of building regulations in some of the more common building situations. They set out what, in ordinary circumstances, may be accepted as reasonable provision for compliance with the relevant requirement(s) of building regulations to which they refer.

1.3 If guidance in an Approved Document is followed there will be a presumption of compliance with the requirement(s) covered by the guidance. However, this presumption can be overturned, so simply following guidance does not guarantee compliance; for example, if the particular case is unusual in some way, then 'normal' guidance may not be applicable. It is also important to note that there may well be other ways of achieving compliance with the requirements. **There is therefore no obligation to adopt any particular solution contained in this Approved Document if you would prefer to meet the relevant requirement in some other way. Persons intending to carry out building work should always check with their building control body, either the local authority or an approved inspector, that their proposals comply with building regulations.**

1.4 It is important to note that this Approved Document, as well as containing guidance, also contains extracts from the Regulations. Such regulatory text must be complied with as stated. For example, the requirement that the **fixed building services** must be commissioned (Regulation 20C) is a regulatory requirement. There is therefore no flexibility to ignore this requirement; neither can compliance with this particular regulation be demonstrated via any route other than that set out in Regulation 20C.

1.5 The guidance contained in this Approved Document relates only to the particular requirements of the Building Regulations that the document addresses (set out in Section 2). However, building work may be subject to more than one requirement of building regulations. In such cases the work will also have to comply with any other applicable requirements of building regulations.

1.6 There are Approved Documents that give guidance on each of the Parts of Schedule 1 and on regulation 7. A full list of these is provided at the back of this document.

Consideration of technical risk

1.7 Building work to existing buildings must satisfy all the technical requirements set out in regulations 4A, 4B, 17D and 17E of, and Schedule 1 to, the Building Regulations. When considering the incorporation of energy efficiency measures in buildings, attention should also be paid in particular to the need to comply with Part B (fire safety), Part C (site preparation and resistance to contaminants and moisture), Part E (resistance to the passage of sound), Part F (ventilation), paragraph G3 (hot water supply and systems), Part J (combustion appliances and fuel storage systems) and Part P (electrical safety) of Schedule 1 to the Building Regulations, as well as Part L. The adoption of any particular energy efficiency measure should not involve unacceptable technical risk of, for instance, excessive condensation. Designers and builders should refer to the relevant Approved Documents and to other generally available good practice guidance to help minimise these risks.

How to use this Approved Document

1.8 This Approved Document is subdivided into seven sections as detailed below. These main sections are followed by supporting appendices.

This **introductory** section sets out the general context in which the guidance in the Approved Document must be considered.

Section 2 sets out the relevant legal requirements contained in the Building Regulations.

Section 3 contains general guidance, including the definition of key terms, the types of building work covered by this Approved Document, the types of building work that are exempt, procedures for notifying work, materials and workmanship and health and safety issues, an overview of the routes to compliance, and how to deal with 'special' areas of buildings that contain **dwellings**.

Section 4 details the considerations that apply to demonstrating that the building work will meet the **energy efficiency requirements**. This section begins the detailed technical guidance relating to showing compliance with the **energy efficiency requirements.**

Section 5 details the considerations that apply when providing or renovating **thermal elements**.

Section 6 gives guidance about the requirement for **consequential improvements** for buildings over 1,000 m^2.

INTRODUCTION

L2B

Section 7 describes the information that should be provided to occupiers to help them achieve reasonable standards of energy efficiency in practice.

1.9 In this document the following conventions have been adopted to assist understanding and interpretation:

a. Texts shown against a green background are extracts from the Building Regulations or Building (Approved Inspectors etc.) Regulations 2000 (SI 2000/2532)('the Approved Inspectors Regulations'), both as amended, and set out the legal requirements that relate to compliance with the ***energy efficiency requirements*** of building regulations. As stated previously, there is no flexibility in respect of such text; it defines a legal requirement, not guidance for typical situations. It should also be remembered that, as noted above, building works must comply with all the other applicable requirements of building regulations.

b. Key terms are defined in paragraph 3.1 and are printed in ***bold italic text.***

c. Details of technical publications referred to in the text of this Approved Document will be given in footnotes and repeated as references at the end of the document. A reference to a publication is likely to be made for one of two main reasons. The publication may contain additional or more comprehensive technical detail, which it would be impractical to include in full in the Approved Document but which is needed to fully explain ways of meeting the requirements; or it is a source of more general information. The reason for the reference will be indicated in each case. The reference will be to a specified edition of the document. The Approved Document may be amended from time to time to include new references or to refer to revised editions where this aids compliance.

d. Additional *commentary in italic text* appears after some numbered paragraphs. This commentary is intended to assist understanding of the immediately preceding paragraph or sub-paragraph, or to direct readers to sources of additional information, but is not part of the technical guidance itself.

Where you can get further help

1.10 If you do not understand the technical guidance or other information set out in this Approved Document and the additional detailed technical references to which it directs you, there are a number of routes through which you can seek further assistance:

- the CLG website: www.communities.gov.uk;
- the Planning Portal website: www.planningportal.gov.uk;
- if you are the person undertaking the building work, you can seek assistance either from your local authority building control service or from your approved inspector (depending on which building control service you are using);
- persons registered with a competent person self-certification scheme may be able to get technical advice from their scheme operator;
- if your query is of a highly technical nature, you may wish to seek the advice of a specialist, or industry technical body, for the relevant subject.

Responsibility for compliance

1.11 It is important to remember that if you are the person (e.g. designer, builder, installer) carrying out building work to which any requirement of building regulations applies you have a responsibility to ensure that the work complies with any such requirement. The building owner may also have a responsibility for ensuring compliance with building regulation requirements and could be served with an enforcement notice in cases of non-compliance.

Section 2: The Requirements

2.1 This Approved Document, which takes effect on 1 October 2010, deals with the *energy efficiency requirements* in the Building Regulations (as amended). Regulation 2(1) of the Building Regulations defines the *energy efficiency requirements* as the requirements of regulations 4A, 17C, 17D and 17E and Part L of Schedule 1. The *energy efficiency requirements* relevant to existing buildings other than *dwellings* are those of regulations 4A, 17D and 17E of, and Part L of Schedule 1 to, those Regulations, as set out below.

Requirements relating to thermal elements – Regulation 4A

(1) Where a person intends to renovate a thermal element, such work shall be carried out as is necessary to ensure that the whole thermal element complies with the requirements of paragraph L1(a)(i) of Schedule 1.

(2) Where a thermal element is replaced, the new thermal element shall comply with the requirements of paragraph L1(a) (i) of Schedule 1.

Consequential improvements to energy performance – Regulation 17D

(1) Paragraph (2) applies to an existing building with a total useful floor area over 1000 m^2 where the proposed building work consists of or includes–

(a) an extension;

(b) the initial provision of any fixed building services; or

(c) an increase to the installed capacity of any fixed building services.

(2) Subject to paragraph (3), where this paragraph applies, such work, if any, shall be carried out as is necessary to ensure that the building complies with the requirements of Part L of Schedule 1.

(3) Nothing in paragraph (2) requires work to be carried out if it is not technically, functionally or economically feasible.

Regulation 17J defines 'building' in regulation 17D as follows:

Regulation 17J Interpretation

In this Part–

'building' means the building as a whole or parts of it that have been designed or altered to be used separately.

Energy performance certificates – Regulation 17E

(1) This regulation applies where–

(a) a building is erected; or

(b) a building is modified so that it has a greater or fewer number of parts designed or altered for separate use than it previously had, where the modification includes the provision or extension of any of the fixed services for heating, hot water, air conditioning or mechanical ventilation.

THE REQUIREMENTS

Energy performance certificates – Regulation 17E *(continued)*

(2) The person carrying out the work shall–

(a) give an energy performance certificate for the building to the owner of the building; and

(b) give to the local authority notice to that effect, including the reference number under which the energy performance certificate has been registered in accordance with regulation 17F(4).

(3) The energy performance certificate and notice shall be given not later than five days after the work has been completed.

(4) The energy performance certificate must be accompanied by a recommendation report containing recommendations for the improvement of the energy performance of the building, issued by the energy assessor who issued the energy performance certificate.

(5) An energy performance certificate must–

(a) express the asset rating of the building in a way approved by the Secretary of State under regulation 17A;

(b) include a reference value such as a current legal standard or benchmark;

(c) be issued by an energy assessor who is accredited to produce energy performance certificates for that category of building; and

(d) include the following information–

(i) the reference number under which the certificate has been registered in accordance with regulation 17F(4);

(ii) the address of the building;

(iii) an estimate of the total useful floor area of the building;

(iv) the name of the energy assessor who issued it;

(v) the name and address of the energy assessor's employer, or, if he is self-employed, the name under which he trades and his address;

(vi) the date on which it was issued; and

(vii) the name of the approved accreditation scheme of which the energy assessor is a member.

(6) Certification for apartments or units designed or altered for separate use in blocks may be based–

(a) except in the case of a dwelling, on a common certification of the whole building for blocks with a common heating system; or

(b) on the assessment of another representative apartment or unit in the same block.

(7) Where–

(a) a block with a common heating system is divided into parts designed or altered for separate use; and

(b) one or more, but not all, of the parts are dwellings,

certification for those parts which are not dwellings may be based on a common certification of all the parts which are not dwellings.

Requirement	Limits on application

Schedule 1 – Part L Conservation of fuel and power

L1. Reasonable provision shall be made for the conservation of fuel and power in buildings by:

(a) limiting heat gains and losses–

(i) through thermal elements and other parts of the building fabric; and

(ii) from pipes, ducts and vessels used for space heating, space cooling and hot water services;

(b) providing fixed building services which–

(i) are energy efficient;

(ii) have effective controls; and

(iii) are commissioned by testing and adjusting as necessary to ensure they use no more fuel and power than is reasonable in the circumstances; and

(c) providing to the owner sufficient information about the building, the fixed building services and their maintenance requirements so that the building can be operated in such a manner as to use no more fuel and power than is reasonable in the circumstances.

LIMITATION ON REQUIREMENTS

2.2 In accordance with regulation 8 of the Building Regulations, the requirements in Parts A to D, F to K, N and P (except for paragraphs G2, H2 and J6) of Schedule 1 to the Building Regulations do not require anything to be done except for the purpose of securing reasonable standards of health and safety for persons in or about buildings (and any others who may be affected by buildings or matters connected with buildings).

2.3 Paragraph G2 is excluded as it deals with water efficiency and paragraphs H2 and J6 are excluded from regulation 8 because they deal directly with prevention of the contamination of water. Parts E and M (which deal, respectively, with resistance to the passage of sound, and access to and use of buildings) are excluded from regulation 8 because they address the welfare and convenience of building users. Part L is excluded from regulation 8 because it addresses the conservation of fuel and power.

Section 3: General guidance

Key terms

3.1 The following are key terms used in this document:

BCB means Building Control Body: a local authority or an approved inspector.

Commissioning means the advancement of a ***fixed building service*** following installation, replacement or alteration of the whole or part of the system, from the state of static completion to working order by testing and adjusting as necessary to ensure that the system as a whole uses no more fuel and power than is reasonable in the circumstances, without prejudice to the need to comply with health and safety requirements. For each system ***commissioning*** includes setting-to-work, regulation (that is testing and adjusting repetitively) to achieve the specified performance, the calibration, setting up and testing of the associated automatic control systems, and recording of the system settings and the performance test results that have been accepted as satisfactory.

Consequential improvements means those energy efficiency improvements required by regulation 17D.

Controlled service or fitting means a service or fitting in relation to which Part G (sanitation, hot water safety and water efficiency), H (drainage and waste disposal), J (combustion appliances and fuel storage systems), L (conservation of fuel and power) or P (electrical safety) of Schedule 1 to the Building Regulations imposes a requirement.

Display window means an area of glazing, including glazed doors, intended for the display of products or services on sale within the building, positioned at the external perimeter of the building, at an access level and immediately adjacent to a pedestrian thoroughfare. There should be no permanent workspace within one glazing height of the perimeter. Glazing that extends beyond 3 m above such an access level is not part of a ***display window*** except:

a. where the products on display require a greater height of glazing;

b. in existing buildings, when replacing ***display windows*** that already extend to a greater height;

c. in cases of building work involving changes to the façade and glazing and requiring planning consent, where planners have discretion to require a greater height of glazing, e.g. to fit in with surrounding buildings or to match the character of the existing façade.

It is expected that ***display windows*** will be found in buildings in Planning Use Classes A1, A2, A3 and D2 as detailed in Table 1.

Table 1 **Planning Use Classes**

Class	Use
A1	Shops: including retail-warehouse, undertakers, showrooms, post offices, hairdressers, shops for sale of cold food for consumption off premises
A2	Financial and professional services: banks, building societies, estate and employment agencies, betting offices
A3	Food and drink: restaurants, pubs, wine bars, shops for sale of hot food for consumption off premises
D2	Assembly and leisure: cinemas, concert halls, bingo halls, casinos, sports and leisure uses

Dwelling means a self-contained unit, including a house or flat, designed to be used separately to accommodate a single household. (***Rooms for residential purposes*** are not ***dwellings*** so this Approved Document L2B applies to work in such buildings).

Emergency escape lighting means that part of emergency lighting that provides illumination for the safety of people leaving an area or attempting to terminate a dangerous process before leaving an area.

Energy efficiency requirements means the requirements of regulations 4A, 17C, 17D and 17E of, and Part L of Schedule 1 to, the Building Regulations.

In respect of existing buildings the applicable requirements consist of Part L and regulations 4A and 17D.

Fit-out work means that work needed to complete the internal layout and servicing of the building shell to meet the specific needs of an incoming occupier. The building shell is the structural and non-structural envelope of a building provided as a primary stage (usually for a speculative developer) for a subsequent project to fit out with internal accommodation works.

Fixed building services means any part of, or any controls associated with:

a. fixed internal or external lighting systems but does not include emergency escape lighting or specialist process lighting; or

b. fixed systems for heating, hot water service, air-conditioning or mechanical ventilation.

High-usage entrance door means a door to an entrance primarily for the use of people that is expected to experience large traffic volumes, and where robustness and/or powered operation is the primary performance requirement. To qualify as a ***high-usage entrance door***, the door should be equipped with automatic closers and, except where operational requirements preclude, be protected by a lobby.

L2B GENERAL GUIDANCE

Principal works means the work necessary to achieve the client's purposes in extending the building and/or increasing the installed capacity of any ***fixed building services***. The value of the ***principal works*** is the basis for determining a reasonable provision of ***consequential improvements***.

Renovation in relation to a thermal element means the provision of a new layer in the thermal element or the replacement of an existing layer, but excludes decorative finishes, and 'renovate' shall be construed accordingly.

Simple payback means the amount of time it will take to recover the initial investment through energy savings, and is calculated by dividing the marginal additional cost of implementing an energy efficiency measure by the value of the annual energy savings achieved by that measure taking no account of VAT. When making this calculation, the following guidance should be used:

a. The marginal additional cost is the additional cost (materials and labour) of incorporating (e.g.) additional insulation, not the whole cost of the work.

b. The cost of implementing the measure should be based on prices current at the date the proposals are made known to the ***BCB*** and be confirmed in a report signed by a suitably qualified person.

c. The annual energy savings should be estimated using an energy calculation tool approved by the Secretary of State pursuant to regulation 17A.

d. For the purposes of this Approved Document, the energy prices that are current at the time of the application to building control should be used when evaluating the annual energy savings. Current energy prices can be obtained from the DECC website[1].

Thermal element is defined in regulation 2(2A) of the Building Regulations as follows:

(2A) In these Regulations 'thermal element' means a wall, floor or roof (but does not include windows, doors, roof windows or roof-lights) which separates a thermally conditioned part of the building ('the conditioned space') from:

a. the external environment (including the ground); or

b. in the case of floors and walls, another part of the building which is:

　i. unconditioned;

　ii. an extension falling within class VII in Schedule 2; or

　iii. where this paragraph applies, conditioned to a different temperature,

and includes all parts of the element between the surface bounding the conditioned space and the external environment or other part of the building as the case may be.

(2B) Paragraph (2A)(b)(iii) applies only to a building which is not a dwelling, where the other part of the building is used for a purpose which is not similar or identical to the purpose for which the conditioned space is used.

Note that this definition encompasses the walls and floor of a swimming pool basin where this is part of an existing dwelling.

Total useful floor area is the total area of all enclosed spaces measured to the internal face of the external walls, that is to say it is the gross floor area as measured in accordance with the guidance issued to surveyors by the RICS. In this convention:

a. the area of sloping surfaces such as staircases, galleries, raked auditoria, and tiered terraces should be taken as their area on plan; and

b. areas that are not enclosed such as open floors, covered ways and balconies are excluded.

This equates to the gross floor area as measured in accordance with the guidance issued to surveyors by the RICS.

Types of work covered by this Approved Document

3.2 This Approved Document is intended to give guidance on what, in ordinary circumstances, may be considered reasonable provision for compliance with the requirements of regulations 4A and 17D of, and Part L of Schedule 1 to, the Building Regulations when carrying out work on existing buildings that are not ***dwellings***. In addition it gives guidance on compliance with regulations 20B, 20C and 20D of the Building Regulations and 12B, 12C and 12D of the Approved Inspectors Regulations.

*It should be noted that **dwellings** are defined as self-contained units. **Rooms for residential purposes** are not **dwellings**, and so this Approved Document applies to them.*

3.3 In particular, this Approved Document gives guidance on compliance with the ***energy efficiency requirements*** where the following occurs:

a. the construction of an extension (see paragraphs 4.1 to 4.13);

b. a material change of use or a change to the building's energy status (paragraphs 4.15 to 4.21);

c. the provision or extension of a ***controlled fitting*** or ***controlled service*** (see paragraphs 4.22 to 4.48;

---

1 http://www.decc.gov.uk/en/content/cms/statistics/publications/prices/prices.aspx

GENERAL GUIDANCE

L2B

d. the replacement or **renovation** of a **thermal element** (Section 5);

e. **consequential improvements** (Section 6).

3.4 For certain types of work in relation to an existing building, it may be more appropriate to use the guidance from the other Approved Documents L, or to follow only a limited amount of the guidance in this Approved Document. The following sub-paragraphs identify some of the circumstances in which this might be appropriate:

a. For <u>first</u> **fit-out works** in buildings such as shell and core office buildings or business park units, the guidance in Approved Document L2A (new non-domestic buildings) covering first fit-out should be followed (but note that the appropriate guidance for any <u>subsequent</u> **fit-out works** is contained in this Approved Document).

b. Large extensions (as defined in paragraph 4.2) should be carried out in accordance with the guidance in Approved Document L2A. However, regulation 17D (**consequential improvements** to energy performance) may apply, in which case the guidance in relation to that regulation set out in this Approved Document would be relevant.

c. Modular and portable buildings: where the work involves the construction of subassemblies that have been obtained from a centrally held stock or from the disassembly or relocation of such buildings at other premises, the guidance in Approved Document L2A should be followed but regulation 17D (**consequential improvements** to energy performance) may also apply if the work was to extend an existing building. In that context, the guidance in relation to that regulation as set out in this Approved Document would be relevant.

Note that erecting a separate unit on a site with an existing building is not extending that existing building, but is the creation of a new building, unless the new unit is to be permanently linked to the existing building.

d. Where the work involves a building that either before the work or after the work is completed contains one or more **dwellings**, the guidance in Approved Document L1B would apply to each **dwelling**.

*It should be noted that **dwellings** are defined as self-contained units. **Rooms for residential purposes** are not **dwellings**, and so this Approved Document applies to them.*

Buildings exempt from the energy efficiency requirements

3.5 Building work in most existing buildings other than **dwellings** will need to comply with the **energy efficiency requirements** of the Building Regulations where the buildings are roofed constructions having walls and use energy to condition the indoor climate. Regulation 9 of the Regulations, however, grants an exemption from compliance with the **energy efficiency requirements** to certain classes of buildings:

a. buildings which are:

  i. listed in accordance with section 1 of the Planning (Listed Buildings and Conservation Areas) Act 1990;

  ii. in a conservation area designated in accordance with section 69 of that Act; or

  iii. included in the schedule of monuments maintained under section 1 of the Ancient Monuments and Archaeological Areas Act 1979,

  where compliance with the **energy efficiency requirements** would unacceptably alter their character or appearance;

b. buildings which are used primarily or solely as places of worship;

c. temporary buildings with a planned time of use of 2 years or less, industrial sites, workshops and non-residential agricultural buildings with low energy demand;

d. stand-alone buildings other than **dwellings** with a **total useful floor area** of less than 50 m^2;

e. carports, covered yards, covered ways and some conservatories or porches attached to existing buildings. Guidance on these is given at paragraphs 3.21 and 3.22 below.

Special considerations

3.6 Special considerations apply to certain classes of non-exempt building. These building types are:

a. historic buildings and buildings used primarily or solely as places of worship; the considerations that apply to such existing buildings are given in paragraphs 3.9 to 3.14;

b. buildings with low energy demand; the guidance specific to such buildings is given in paragraphs 3.15 to 3.20;

c. modular and portable buildings; for the construction of such buildings with a planned time of use of more than 2 years at one or more locations, the guidance in Approved Document L2A should be followed. Any changes to the building fabric or **fixed building services** should comply with this Approved Document.

Historic and traditional buildings which may have an exemption

3.7 As mentioned above in paragraph 3.5 the following classes of buildings have an exemption from the **energy efficiency requirements** <u>where compliance would unacceptably alter the character or appearance of the buildings:</u>

a. listed buildings;

b. buildings in conservation areas; and

c. scheduled ancient monuments.

L2B GENERAL GUIDANCE

Historic and traditional buildings where special considerations may apply

3.8 There are three further classes of buildings where special considerations in making reasonable provision for the conservation of fuel or power may apply:

a. buildings which are of architectural and historical interest and which are referred to as a material consideration in a local authority's development plan or local development framework;

b. buildings which are of architectural and historical interest within national parks, areas of outstanding natural beauty, registered historic parks and gardens, registered battlefields, the curtilages of scheduled ancient monuments and world heritage sites;

c. buildings of traditional construction with permeable fabric that both absorbs and readily allows the evaporation of moisture.

3.9 When undertaking work on or in connection with a building that falls within one of the classes listed above, the aim should be to improve energy efficiency as far as is reasonably practical. The work should not prejudice the character of the host building or increase the risk of long-term deterioration of the building fabric or fittings.

3.10 The guidance given by English Heritage[2] should be taken into account in determining appropriate energy performance standards for building work in historic buildings.

3.11 In general, new extensions to historic or traditional buildings should comply with the standards of energy efficiency as set out in this Approved Document. The only exception would be where there is a particular need to match the external appearance or character of the extension to that of the host building.

3.12 Particular issues relating to work in historic buildings that warrant sympathetic treatment and where advice from others could therefore be beneficial include:

a. restoring the historic character of a building that has been subject to previous inappropriate alteration, e.g. replacement windows, doors and rooflights;

b. rebuilding a former historic building (e.g. following a fire or filling a gap site in a terrace);

c. making provisions enabling the fabric of historic buildings to 'breathe' to control moisture and potential long-term decay problems.

3.13 In assessing reasonable provision for energy efficiency improvements for historic buildings of the sort described in paragraphs 3.7 and 3.8, it is important that the **BCB** takes into account the advice of the local authority's conservation officer. The views of the conservation officer are particularly important where building work requires planning permission and/or listed building consent.

Places of worship

3.14 For the purposes of the *energy efficiency requirements*, places of worship are taken to mean those buildings or parts of a building that are used for formal public worship, including adjoining spaces whose function is directly linked to that use. Such parts of buildings of this type often have traditional, religious or cultural constraints that mean that compliance with the *energy efficiency requirements* would not be possible. Other parts of the building that are designed to be used separately, such as offices, catering facilities, day centres and meeting halls are not exempt.

Industrial sites, workshops and non-residential agricultural buildings with low energy demand

3.15 In relation to this category of exempt building, the low energy demand relates only to the energy used by fixed heating or cooling systems, NOT to energy required for or created by process needs. The following are examples of buildings in the above categories that are low energy demand:

a. buildings or parts of buildings where the space is not generally heated, other than by process heat, or cooled;

b. buildings or parts of buildings that require heating or cooling only for short periods each year, such as during critical periods in the production cycle (e.g. plant germination, egg hatching) or in very severe weather conditions.

3.16 Industrial sites, workshops and non-residential agricultural buildings are only exempt if they meet the low energy demand criterion. If not exempt, such buildings must comply with *energy efficiency requirements*. Similarly, other buildings (e.g. some types of warehouse) may have low energy demand but are not exempt because they do not fall into one of the above categories.

Non-exempt buildings with low energy demand

3.17 For the purposes of this Approved Document, non-exempt buildings with low energy demand are taken to be those buildings or parts thereof where:

a. *fixed building services* are used to heat or cool only a localised area rather than the entire enclosed volume of the space concerned (e.g. localised radiant heaters at a workstation in a generally unheated space); or

---

[2] www.english-heritage.org.uk

GENERAL GUIDANCE

b. *fixed building services* are used to heat spaces in the building to temperatures substantially lower than those normally provided for human comfort (e.g. to provide condensation or frost protection in a warehouse).

3.18 In such situations, it is not reasonable to expect the entire building envelope to be insulated to the standard expected for more typical buildings. Therefore, if an existing building with low levels of heating is extended or parts of the fabric renovated, the new or renovated building envelope should be insulated only to a degree that is reasonable in the particular case. If some general heating is provided (case b above), then it would be reasonable that no part of the opaque fabric had a U-value worse than 0.7 $W/m^2.K$. In addition, reasonable provision would be for every *fixed building service* that is installed to meet the energy efficiency standards set out in the *Non-Domestic Building Services Compliance Guide*[3].

3.19 If a part of a building with low energy demand is partitioned off and heated normally (e.g. an office area in an unheated warehouse), the separately heated area should be treated as a separate 'building' and the normal procedures for demonstrating compliance should be followed in respect of the enclosure.

3.20 If a building with low energy demand subsequently changes such that the space is generally conditioned, then this is likely to involve the initial provision or an increase in the installed capacity of a *fixed building service*. Such activities may fall within regulation 17D, which would require the building envelope to be upgraded and *consequential improvements* to be made (see the guidance in Section 6 of this Approved Document). Alternatively, if the building shell was designed as a building with low energy demand and the first occupier of the building wanted to install (e.g.) heating, this would be a first *fit-out works*, and Approved Document L2A would apply.

Conservatories and porches

3.21 Regulation 9 of the Building Regulations exempts some conservatory and porch extensions from the *energy efficiency requirements*. The exemption applies only to conservatories or porches:

- which are at ground level;
- where the floor area is less than $30m^2$;
- where the existing walls, doors and windows which separate the conservatory from the building are retained or, if removed, are replaced by walls, windows and doors which meet the *energy efficiency requirements*; and
- where the heating system of the building is not extended into the conservatory or porch.

3.22 Where any conservatory or porch does not meet all the requirements in the preceding paragraph, it is not exempt and must comply with the relevant *energy efficiency requirements* (see paragraphs 4.12 and 4.13 below).

Notification of work covered by the energy efficiency requirements

3.23 In most instances, in order to comply with the Building Regulations it will be necessary to notify a **BCB** before the work starts. If you choose to use the local authority this must be by deposit of full plans. There is no set procedure where the **BCB** is an Approved Inspector provided it has been notified at least 5 days before work starting.

3.24 In certain situations, however, other procedures apply:

a. Where the work is being carried out by a person registered with a relevant competent person self-certification scheme listed in Schedule 2A to the Building Regulations, no advance notification to the **BCB** is needed (see paragraphs 3.25 to 3.28);

b. Where the work involves an emergency repair, e.g. to a failed boiler or a leaking hot water cylinder, in accordance with regulation 12(7) of the Building Regulations there is no need to delay making the repair in order to make an advance notification to the **BCB**. However, in such cases it will still be necessary for the work to comply with the relevant requirements and to give a notice to the **BCB** at the earliest opportunity, unless an installer registered under an appropriate competent person scheme carries out the work. A completion certificate can then be issued in the normal way;

c. Where the work is of a minor nature as described in the schedule of non-notifiable work (Schedule 2B to the Building Regulations), the work must still comply with the relevant requirements but need not be notified to the **BCB** (see paragraphs 3.29 and 3.30).

Competent person self-certification schemes

3.25 It is not necessary to notify a **BCB** in advance of work which is to be carried out by a person registered with a relevant competent person self-certification scheme listed in Schedule 2A to the Building Regulations. In order to join such a scheme, a person must demonstrate competence to carry out the type of work the scheme covers, and also the ability to comply with all relevant requirements in the Building Regulations.

---

[3] *Non-Domestic Building Services Compliance Guide*, CLG, 2010.

L2B GENERAL GUIDANCE

3.26 Where work is carried out by a person registered with a competent person scheme, regulation 16A of the Building Regulations and regulation 11A of the Approved Inspectors Regulations require that the occupier of the building be given, within 30 days of the completion of the work, a certificate confirming that the work complies fully with all applicable building regulation requirements. There is also a requirement to give the **BCB** a notice of the work carried out, again within 30 days of the completion of the work. These certificates and notices are usually made available through the scheme operator.

3.27 **BCBs** are authorised to accept these certificates and notices as evidence of compliance with the requirements of the Building Regulations. Local authority inspection and enforcement powers remain unaffected, although they are normally used only in response to a complaint that work does not comply.

3.28 A list of authorised self-certification schemes and the types of work for which they are authorised can be found at www.communities.gov.uk.

Work which need not be notified

3.29 Schedule 2B to the Building Regulations sets out types of work where there is no requirement to notify a **BCB** that work is to be carried out. These types of work are mainly of a minor nature where there is no significant risk to health, safety or energy efficiency. Note that the health, safety and **energy efficiency requirements** continue to apply to these types of work, and that only the need to notify a **BCB** has been removed. In addition, where only non-notifiable work is carried out by a member of a competent person self-certification scheme there is no requirement for a certificate of building regulations compliance to be given to the occupier or the **BCB**.

3.30 The types of non-notifiable work in Schedule 2B relevant to the **energy efficiency requirements** of the Regulations are:

a. In a heating, hot water service, ventilation or air-conditioning system, the replacement of any part which is not a combustion appliance (such as a radiator, valve or pump) or the addition of an output device (such as a radiator or fan) or the addition of a control device (such as a thermostatic radiator valve). However, the work will remain notifiable whenever **commissioning** is possible and necessary to enable a reasonable use of fuel and power.

b. The installation of a stand-alone, self-contained fixed heating, hot water, ventilation or air-conditioning service. Such services must consist only of a single appliance and any associated controls, and must not be connected to any other **fixed building** service. Examples of non-notifiable services would be a fixed electric heater, a mechanical extractor fan in a kitchen or bathroom, and a room air-conditioning unit. However, if any of the following apply, the work will remain notifiable building work:

   i. the service is a combustion appliance; or

   ii. **commissioning** is possible and would affect the service's energy efficiency (see paragraphs 4.36 to 4.48); or

   iii. in the case of a ventilation appliance, the appliance is installed in a room containing a natural draught open-flued combustion appliance or service, such as a gas fire which uses a chimney as its flue.

c. Installation of thermal insulation in a roof space or loft space where this is the only work carried out and the work is carried out voluntarily and not in order to comply with any requirement in the Building Regulations.

Materials and workmanship

3.31 Any building work which is subject to the requirements imposed by Schedule 1 to the Building Regulations should, in accordance with regulation 7, be carried out with proper materials and in a workmanlike manner.

3.32 You may show that you have complied with regulation 7 in a number of ways. These include demonstrating the appropriate use of:

- a product bearing CE marking in accordance with the Construction Products Directive (89/106/EEC)[4] as amended by the CE Marking Directive (93/68/EC)[5], the Low Voltage Directive (2006/95/EC)[6] and the EMC Directive (2004/108/EC)[7];

- a product complying with an appropriate technical specification (as defined in those Directives mentioned above), a British Standard, or an alternative national technical specification of a Member State of the European Union or Turkey[8], or of another State signatory to the Agreement on the European Economic Area (EEA) that provides an equivalent level of safety and protection;

- a product covered by a national or European certificate issued by a European Technical Approval Issuing body, provided the conditions of use are in accordance with the terms of the certificate.

3.33 You will find further guidance in the Approved Document which specifically supports regulation 7 on materials and workmanship.

---

[4] As implemented by the Construction Products Regulations 1991 (SI 1991/1620).
[5] As implemented by the Construction Products (Amendment) Regulations 1994 (SI 1994/3051).
[6] As implemented by the Electrical Equipment (Safety) Regulations 1994 (SI 1994/3260).
[7] As implemented by the Electromagnetic Compatibility Regulations 2006 (SI 2006/3418).
[8] Decision No 1/95 of the EC-Turkey Association Council of 22 December 1995.

GENERAL GUIDANCE

Independent certification schemes

3.34 There are many UK product certification schemes. Such schemes certify compliance with the requirements of a recognised standard that is appropriate to the purpose for which the material is to be used. Materials which are not so certified may still conform to a relevant standard.

3.35 Many certification bodies that approve products under such schemes are accredited by the United Kingdom Accreditation Service (UKAS). Such bodies can issue certificates only for the categories of product covered under the terms of their accreditation.

3.36 **BCBs** may take into account the certification of products, components, materials or structures under such schemes in deciding whether compliance with the relevant standard has been achieved. Similarly, **BCBs** may take account of the certification of the installation or maintenance of products, components, materials or structures under such schemes in deciding on compliance with the relevant standard. Nonetheless, before accepting that certification constitutes compliance with building regulations, a **BCB** should establish in advance that the relevant scheme is adequate for that purpose.

Standards and technical specifications

3.37 Building regulations are made for specific purposes including securing the health, safety, welfare and convenience of people in or about buildings; furthering the conservation of fuel and power; furthering the protection or enhancement of the environment; and facilitating sustainable development. Guidance contained in standards and technical approvals referred to in Approved Documents may be relevant to compliance with building regulations to the extent that it relates to those purposes. However, it should be noted that guidance in standards and technical approvals may also address other aspects of performance such as serviceability, or aspects which, although they relate to health and safety, are not covered by building regulations.

3.38 When an Approved Document makes reference to a named standard or document, the relevant version of the standard or document is the one listed at the end of the Approved Document. Until the reference in the Approved Document is revised, the standard or document listed remains the approved source, but if the issuing body has published a revised or updated version, any content that addresses the relevant requirements of the Building Regulations may be used as a source of guidance.

3.39 The appropriate use of a product that complies with a European Technical Approval as defined in the Construction Products Directive will meet the relevant requirements.

3.40 Communities and Local Government intends to issue periodic amendments to its Approved Documents to reflect emerging harmonised European Standards. Where a national standard is to be replaced by a European harmonised standard, there will be a coexistence period during which either standard may be referred to. At the end of the coexistence period the national standard will be withdrawn.

The Workplace (Health, Safety and Welfare) Regulations 1992

3.41 The Workplace (Health, Safety and Welfare) Regulations 1992, as amended, apply to the common parts of flats and similar buildings if people such as cleaners, wardens and caretakers are employed to work in these common parts. These Regulations contain some requirements which affect building design. The main requirements are now covered by the Building Regulations, but for further information see *Workplace health, safety and welfare, Workplace (Health, Safety and Welfare) Regulations 1992, Approved Code of Practice and guidance,* HSE publication L24, 1996.

Section 4: Guidance relating to building work

EXTENSIONS

4.1 Under regulation 17D of the Building Regulations, the construction of an extension triggers the requirement for *consequential improvements* in buildings with a *total useful floor area* greater than 1000m². In such cases, the guidance in Section 6 should be followed in addition to the following specific guidance.

Large extensions

4.2 Where the proposed extension has a *total useful floor area* that is both:

a. greater than 100 m², and

b. greater than 25 per cent of the *total useful floor area* of the existing building,

the work should be regarded as a new building and the guidance in Approved Document L2A followed. The requirement for *consequential improvements*, if appropriate, should also be met by following the guidance in Section 6 of this Approved Document.

Other extensions – reference method

Fabric standards

4.3 Reasonable provision would be for the proposed extension to incorporate the following:

a. doors, windows, roof windows, rooflights and smoke vents that meet the standards set out in paragraphs 4.21 to 4.28;

b. newly constructed *thermal elements* that meet the standards set out in paragraphs 5.1 to 5.7;

c. existing opaque fabric which becomes a *thermal element* where previously it was not should be upgraded so that it meets the standards in paragraphs 5.12 to 5.14.

Opening areas

4.4 The area of windows and rooflights in the extension should generally not exceed the values given in Table 2. However, where a greater proportion of glazing is present in the part of the building to which the extension is attached, reasonable provision would be to limit the proportion of glazing in the extension so that it is no greater than the proportion that exists in the part of the building to which it is attached.

Building services systems in the extension

4.5 Where *fixed building services* are provided or extended as part of constructing the extension, reasonable provision would be to follow the guidance in paragraphs 4.29 to 4.48.

Optional approaches with more design flexibility

4.6 The approach set out in paragraphs 4.3 to 4.5 is somewhat prescriptive. The following paragraphs offer more flexible approaches to demonstrating that reasonable provision has been made. These alternative approaches allow some elements of the design to be relaxed through compensating measures elsewhere.

Area-weighted U-value method

4.7 The U-values given in paragraph 4.3 and the opening areas given in paragraph 4.4 may be varied provided that the area-weighted U-value of all the elements in the extension is no greater than that of an extension of the same size and shape that complies with the U-value standards referred to in paragraph 4.3 and the opening area standards in paragraph 4.4. Any *fixed building service* provided or extended as part of constructing the extension should follow the guidance in paragraphs 4.29 to 4.48.

4.8 The area-weighted U-value is given by the following expression:

$$\{(U_1 \times A_1) + (U_2 \times A_2) + (U_3 \times A_3) + \ldots)\} \div \{(A_1 + A_2 + A_3 + \ldots)\}$$

Table 2 Opening areas in the extension

Building type	Windows and personnel doors as % of exposed wall	Rooflights as % of area of roof
Residential buildings where people temporarily or permanently reside	30	20
Places of assembly, offices and shops	40	20
Industrial and storage buildings	15	20
Vehicle access doors and display windows and similar glazing	As required	N/A
Smoke vents	N/A	As required

GUIDANCE RELATING TO BUILDING WORK

Whole building calculation method

4.9 Where even greater design flexibility is required, reasonable provision would be to use an approved calculation tool to demonstrate that the calculated CO_2 emissions from the building and proposed extension are no greater than for the building plus a notional extension complying with the standards of paragraphs 4.3 to 4.5.

Approved Document C gives limiting values for individual elements to minimise condensation risk.

4.10 The specification of the existing building used in conjunction with the notional extension as the basis of setting the CO_2 target for the building work shall include all upgrades that will be included in fulfilment of the requirement for **consequential improvements** (see Section 6).

Otherwise all the low-cost measures would have been taken by the compensatory measures, leaving little leeway for overall improvement.

4.11 Where additional upgrades over and above the **consequential improvements** are proposed in the actual building to compensate for lower performance in the extension, then such upgrades should be implemented to a standard that is no worse than set out in the relevant guidance contained in this Approved Document. The relevant standards for upgrading retained **thermal elements** are as set out in column (b) of Table 5.

Where it is proposed to upgrade, the standards set out in this Approved Document are cost-effective and should be implemented in full. It will be worthwhile implementing them even if the improvement is greater than necessary to achieve compliance. In some cases, therefore, the standard of the extended building may be better than that required by paragraphs 4.1 to 4.10.

Conservatories and porches

4.12 Where the extension is a conservatory or porch that is not exempt from the **energy efficiency requirements** (see paragraphs 3.22 and 3.23 above), then reasonable provision would be to provide:

a. Effective thermal separation between the heated area in the existing building, i.e. the walls, doors and windows between the building and the extension should be insulated and draughtproofed to at least the same extent as in the existing building.

b. Independent temperature and on/off controls to any heating system installed within the extension. Any fixed building service installed within the extension should also conform to the standards set out in paragraphs 4.29 to 4.48.

c. Glazed elements should meet the standards set out in Table 3 and opaque elements should meet the standards set out in Table 4. However, the limitations on total area of windows, roof windows and doors as set out at paragraph 4.4 above do not apply.

4.13 Removing, and not replacing, any of the thermal separation between the building and an existing exempt extension, or extending the building's heating system into the extension, means that the extension ceases to be exempt (see paragraphs 3.21 and 3.22 above). This constitutes a change to the building's energy status (Regulation 4B). In such situations, the extension should be treated as a conventional extension and reasonable provision would be to demonstrate that the extension meets the guidance set out in paragraphs 4.1 to 4.11 above.

Swimming pool basins

4.14 Where a swimming pool is being provided in a building, the U-value of the basin (walls and floor) should be not worse than 0.25 W/m².K as calculated according to BS EN ISO 13370[9].

MATERIAL CHANGE OF USE AND CHANGE OF ENERGY STATUS

Material change of use

4.15 Material changes of use (see regulation 5 of the Building Regulations) covered by this document are where, after the change:

a. the building is used as a hotel or a boarding house, where previously it was not;

b. the building is used as an institution, where previously it was not;

c. the building is used as a public building, where previously it was not;

d. the building is not a building described in Classes I to VI in Schedule 2, where previously it was;

e. the building contains a room for residential purposes, where previously it did not;

f. the building, which contains at least one room for residential purposes, contains a greater or lesser number of such rooms than it did previously; or

g. the building is used as a shop where previously it was not.

Change of energy status

4.16 A change to a building's energy status is defined in regulation 2(1) as:

any change which results in a building becoming a building to which the energy efficiency requirements of these Regulations apply, where previously it was not.

---

[9] BS EN ISO 13370 *Thermal performance of buildings. Heat transfer via the ground. Calculation methods.*

GUIDANCE RELATING TO BUILDING WORK

4.17 The requirements relating to a change to energy status are in regulation 4B(1):

Where there is a change in a building's energy status, such work, if any, shall be carried out to ensure that the building complies with the applicable requirements of Part L of Schedule 1.

4.18 In this regulation 'building' means the building as a whole or parts of the building that have been designed or altered to be used separately.

For example, this could occur where a previously unheated building, or parts of the building that have been designed or altered to be used separately, were to be heated in future, or where a previously exempt building were no longer within the exempted categories. A material alteration (regulation 3(2) and (3)) may result in a change in a building's energy status.

4.19 In normal circumstances, reasonable provision when there is a material change of use or a change to the building's energy status would be:

a. Where **controlled services or fittings** are being provided or extended, to meet the standards set out in paragraphs 4.22 to 4.48. If the area of openings in the newly created building is more than 25 per cent of the total floor area, the area of openings should either be reduced to be not greater than 25 per cent, or the larger area should be compensated for in some other way using the procedure described in paragraph 4.21.

b. Where the work involves the provision of a **thermal element**, to meet the standards set out in paragraphs 5.1 to 5.7.

For the purposes of the Building Regulations, provision means both new and replacement elements.

c. Where any **thermal element** is being retained, to upgrade it following the guidance given in paragraphs 5.12 to 5.14. This guidance should also be followed in respect of any existing element that becomes part of the thermal envelope of the building where previously it was not.

As an example, this would include the party wall between units in a terrace of industrial units which originally were unheated, but heating is to be provided to one of the units.

d. Where an existing window (including roof window or rooflight) or door which separates a conditioned space from an unconditioned space or the external environment has a U-value that is worse than 3.3 W/m^2.K, to follow the guidance in paragraphs 4.23 to 4.28 unless it is a **display window** or **high usage entrance door**. It would be reasonable in these latter cases to make some lesser provision for energy efficiency.

4.20 As well as satisfying the **energy efficiency requirements** in respect of the material change of use or change in energy status, such building work may be one of the triggers for **consequential improvements** – see regulation 17D and Section 6.

Option providing more design flexibility

4.21 To provide more design flexibility, an approved calculation tool can be used to demonstrate that the CO_2 emissions from the building as it will become are no worse than if the building had been improved following the guidance set out in paragraph 4.19.

WORK ON CONTROLLED FITTINGS AND SERVICES

4.22 A **controlled service or fitting** is defined in Regulation 2(1) as follows:

'Controlled service or fitting' means a service or fitting in relation to which Part G, H, J, L or P of Schedule 1 imposes a requirement;

Controlled fittings

4.23 In the context of this Approved Document, the application of the term **controlled fitting** to a window, roof window, rooflight or door refers to a whole unit, i.e. including the frame. Consequently, replacing the glazing whilst retaining an existing frame is not providing a **controlled fitting**, and so such work is not notifiable and does not have to meet the Part L standards, although where practical it would be sensible to do so. Similar arguments apply to a new door in an existing frame.

4.24 Where windows, roof windows, rooflights or doors are to be provided, reasonable provision would be draughtproofed units whose area-weighted average performance is no worse than that given in Table 3. In addition, insulated cavity closers should be installed where appropriate. If a window, pedestrian door or rooflight is enlarged or a new one created, then the area of windows and pedestrian doors and of rooflights expressed as a percentage of the total floor area of the building should not exceed the relevant value from Table 2, or should be compensated for in some other way. Where the replacement windows are unable to meet the requirements of Table 3 because of the need to maintain the external appearance of the façade or the character of the building, replacement windows should meet a centre pane U-value of 1.2 W/m^{2}K, or single glazing should be supplemented with low-e secondary glazing. In this latter case, the weather stripping should be on the secondary glazing to minimise condensation risk between the primary and secondary glazing.

4.25 U-values of windows, roof-windows, rooflights and doors shall be calculated using the methods and conventions set out in BR 443[10], and should be based on the whole unit (i.e. in the case of a window, the combined performance of the glazing and frame). The U-value for windows can be taken as that for:

---

[10] BR 443 *Conventions for U-value calculations*, BRE, 2006

GUIDANCE RELATING TO BUILDING WORK　　L2B

a. the smaller of the two standard windows defined in BS EN 14351-1[11]; or
b. the standard configuration referred to in BR 443; or
c. the specific size and configuration of the actual window.

For domestic type construction, SAP 2009 Table 6e gives values for different window configurations that can be used in the absence of test data or calculated values.

4.26 The U-values for roof windows and rooflights given in this Approved Document are based on the particular U-value having been assessed with the roof window or rooflight in the vertical position. If a particular unit has been assessed in a plane other than the vertical, the standards given in this Approved Document should be modified by making a U-value adjustment following the guidance given in BR 443.

The stated standard for a replacement plastic rooflight as given in Table 3 is 1.8 W/m².K. This is for the unit assessed in the vertical plane. If the performance of a triple-skin rooflight was assessed in the horizontal plane, then, based on the guidance given in BR 443, the standard would be adjusted by 0.3 W/m².K (the value from BR 443 for a horizontal triple-skin rooflight), requiring the rooflight as assessed in the horizontal plane to achieve a standard of 1.8 + 0.3 = 2.1 W/m².K.

4.27 In certain classes of building with high internal gains, a less demanding U-value for glazing may be an appropriate way of reducing overall CO_2 emissions. If this case can be made, then the average U-value for windows, doors and rooflights can be relaxed from the values given in Table 3, but the value should not exceed 2.7 W/m².K.

4.28 The overall U-value of curtain walling should be no greater than the better of 1.8 W/m²K or a limiting U-value U_{limit} given by:

$$U_{limit} = 0.8 + \{(1.2 + (FOL \times 0.5)) \times GF\}$$

where FOL is the fraction of opening lights and GF is the glazed fraction.

This means that if an area of curtain walling is to be 60 per cent glazed and 40 per cent opaque, with 50 per cent opening lights, the U-value standard should be 0.8 + (1.2 + 0.5 × 0.5) × 0.6 = 1.7 W/m².K.

Controlled services

4.29 Where the work involves the provision or extension of **controlled services**, reasonable provision would be demonstrated by following the guidance set out in the *Non-Domestic Building Services Compliance Guide*. The Guide covers the following services:

a. heating and hot water systems (including insulation of pipes, ducts and vessels;
b. mechanical ventilation;

Table 3 Standards for controlled fittings

Fitting	Standard
Windows, roof windows and glazed rooflights[1]	1.8 W/m².K for the whole unit
Alternative option for windows in buildings that are essentially domestic in character[2]	A window energy rating[3] of Band C
Plastic rooflight[4]	1.8 W/m².K
Curtain walling	See paragraph 4.28
Pedestrian doors where the door has more than 50% of its internal face area glazed	1.8 W/m².K for the whole unit
High-usage entrance doors for people	3.5 W/m².K
Vehicle access and similar large doors	1.5 W/m².K
Other doors	1.8 W/m².K
Roof ventilators (including smoke extract ventilators)	3.5 W/m².K

Notes:
1. Display windows are not required to meet the standard given in this table.
2. For example, student accommodation, care homes and similar uses where the occupancy levels and internal gains are essentially domestic in character.
3. See Approved Document L1B for more detail on window energy rating.
4. The relevant rooflight U-value for checking against these limits is that based on the developed area of the rooflight, not the area of the roof aperture.

---

[11] EN 14351-1, *Windows and doors – Product standard, performance characteristics*, 2006.

L2B GUIDANCE RELATING TO BUILDING WORK

c. mechanical cooling/air-conditioning;

d. fixed internal lighting; note that as detailed in Schedule 2B, the work is not notifiable if the floor area that is to be provided with new fixed lighting is not greater than 100m^2. Although not notifiable, the work should still meet the standards set out in the compliance guide.

e. renewable energy systems.

4.30 In general terms, the aim should be to:

a. provide new **fixed building services** that meet reasonable standards of energy efficiency, which in normal circumstances would be:

   i. an efficiency not less than set out in the *Non-Domestic Building Services Compliance Guide*. The efficiency claimed for the **fixed building service** should be based on the appropriate test standard as set out in the Guide and the test data should be certified by a notified body. It would be reasonable for **BCBs** to accept such data at face value. In the absence of such quality assured data, **BCBs** should satisfy themselves that the claimed performance is justified. If a particular technology is not covered in the Guide, reasonable provision would be demonstrated by showing that the proposed technology gives a performance that is no worse than a reference system of the same type whose details are given in the Guide; and

   ii. an efficiency not less than that of the **controlled service** being replaced. If the new service uses a different fuel, the efficiency of the new appliance should be multiplied by the ratio of the CO_2 emission factor of the fuel used in the appliance being replaced to that of the fuel used in the new appliance before making this check.

This will prevent an existing low-carbon component being replaced by a lesser provision when fuel switching. For example, if an existing electric chiller with a Co-efficient of Performance (CoP) of 2.5 is replaced by an absorption chiller with a CoP of 0.8 but fired by waste heat, the equivalent efficiency of the absorption chiller would be 0.8 x (0.517/0.058)=7.1, and so test (ii) would be satisfied. 0.517 and 0.058 kgCO$_2$/kWh are the emission factors for electricity and waste heat respectively[12].

b. provide new HVAC systems with appropriate controls to achieve reasonable standards of energy efficiency. In normal circumstances reasonable provision would be to provide the following control features on each system in addition to the system-specific controls detailed in subsequent paragraphs:

   i. the **fixed building services** systems should be sub-divided into separate control zones to correspond to each area of the building that has a significantly different solar exposure, occupancy period, or type of use;

   ii. each separate control zone should be capable of independent switching and control of set-point;

   iii. the provision of the service should respond to the requirements of the space it serves. If both heating and cooling are provided, they should be controlled so they do not operate simultaneously;

   iv. central plant serving the zone-based systems should operate only as and when required. The default condition should be off;

   v. in addition to these general control requirements, the systems should meet the specific control requirements and general energy efficiency criteria as set out in the *Non-Domestic Building Services Compliance Guide*.

c. demonstrate the new service has been effectively commissioned (see paragraphs 4.36 to 4.48);

d. demonstrate that reasonable provision of energy meters has been made for effective monitoring of the performance of newly installed plant (see paragraphs 4.33 to 4.35);

e. demonstrate that the relevant information has been recorded in a new log book or incorporated into an update of the existing one as described in Section 7.

4.31 If a renewable energy generator such as a wind turbine or photovoltaic array is being replaced, the new system should have an electrical output that is not less than the original installation.

4.32 When replacing a heating appliance, consideration should be given to connecting to any existing local heat networks. If the work involves pipework changes, consideration should be given to providing capped off connections to facilitate subsequent connection to a planned local heat network.

Energy meters

4.33 The aim for buildings as a whole is to enable building occupiers to assign at least 90 per cent of the estimated annual energy consumption of each fuel to the various end-use categories (heating, lighting, etc.).

4.34 Reasonable provision for energy meters in existing buildings would be to install energy metering systems in the building service systems provided as part of the works in accordance with the recommendations in CIBSE TM 39[13].

---

[12] See Table 12 at www.bre.co.uk/sap2009

[13] TM 39 *Building energy metering*, CIBSE, 2010.

GUIDANCE RELATING TO BUILDING WORK

4.35 In addition to this:

a. meters should be provided to enable the performance of any renewable energy system provided as part of the works to be separately monitored;

b. in buildings with a *total useful floor area* greater than 1000 m^2, the metering system should enable automatic meter reading and data collection;

c. the metering provisions should be designed such as to facilitate the benchmarking of energy performance as set out in TM 46[14].

Following implementation of the Energy Services Directive, there are likely to be legal obligations for persons *commissioning* building work on existing buildings with floor areas in excess of 1000 m^2 to notify their intentions to the energy supply companies.

COMMISSIONING OF FIXED BUILDING SERVICES

4.36 Regulation 20C (*Commissioning*) states:

20C–(A1) This regulation applies to building work in relation to which paragraph F1(2) of Schedule 1 imposes a requirement, but does not apply to the provision or extension of any fixed system for mechanical ventilation or any associated controls where testing and adjustment is not possible.

(1) This regulation applies to building work in relation to which paragraph L1(b) of Schedule 1 imposes a requirement, but does not apply to the provision or extension of any fixed building service where testing and adjustment is not possible or would not affect the energy efficiency of that fixed building service.

(2) Where this regulation applies the person carrying out the work shall, for the purpose of ensuring compliance with paragraph F1(2) or L1(b) of Schedule 1, give to the local authority a notice confirming that the fixed building services have been commissioned in accordance with a procedure approved by the Secretary of State.

(3) The notice shall be given to the local authority—

(a) not later than the date on which the notice required by regulation 15(4) is required to be given; or

(b) where that regulation does not apply, not more than 30 days after completion of the work.

4.37 Reasonable provision could be to prepare a *commissioning* plan, identifying the systems that need to be tested and the tests that will be carried out. The notice required by regulation 20C should confirm that the *commissioning* plan has been followed and that every system has been inspected in an appropriate sequence and to a reasonable standard and that the test results confirm that performance is reasonably in accordance with the design requirements.

4.38 Not all *fixed building services* will need to be commissioned. With some systems it is not possible as the only controls are 'on' and 'off' switches. Examples of this would be some mechanical extraction systems or single fixed electrical heaters. In other cases *commissioning* would be possible but in the specific circumstances would have no effect on energy use.

Fixed building services which do not require *commissioning* should be identified in the *commissioning* plan, along with the reason for not requiring *commissioning*.

4.39 *Commissioning* must be carried out in such a way as not to prejudice compliance with any applicable health and safety requirements.

4.40 In existing buildings other than *dwellings commissioning* is most often carried out by the person who installs the system. Sometimes it may be carried out by a subcontractor or by a specialist firm. It is important that whoever carries it out follows the relevant approved procedure.

Notice of completion of commissioning

4.41 The Building Regulations (regulation 20C(2)) and the Building (Approved Inspectors etc) Regulations (regulation 12C(2)) require that a notice be given to the relevant *BCB* that *commissioning* has been carried out according to a procedure approved by the Secretary of State.

4.42 The procedure approved by the Secretary of State is set out in:

a. CIBSE Commissioning Code M on commissioning management[15]; and

This provides guidance on the overall process and includes a schedule of all the relevant guidance documents relating to the ***commissioning*** *of specific building services systems.*

b. for leakage testing of ductwork, paragraphs 4.47 and 4.48.

4.43 Where a building notice or full plans have been given to a *BCB*, the notice should be given within 5 days of the completion of the commissioning work. In other cases, for example where work is carried out by a person registered with a competent person scheme, it must be given within 30 days.

---

[14] TM 46 *Energy benchmarks*, CIBSE, 2008.

[15] CIBSE Commissioning Code M: *Commissioning management*, CIBSE, 2003.

L2B GUIDANCE RELATING TO BUILDING WORK

4.44 Where an approved inspector is the **BCB**, the notice should generally be given within 5 days of the completion of the *commissioning* work. However, where the work is carried out by a person registered with a competent person scheme (see paragraphs 3.25 to 3.28) the notice must be given within 30 days.

4.45 Where the installation of *fixed building services* which require *commissioning* is carried out by a person registered with a competent person scheme the notice of *commissioning* will be given by that person.

4.46 Until the **BCB** receives the *commissioning* notices it may be unable to be reasonably satisfied that Part L has been complied with and consequently may be unable to give a completion/final certificate.

Membership of the Commissioning Specialists Association or the Commissioning group of the HVCA may be a way of demonstrating suitability to sign the report in respect of the HVAC systems. For lighting control systems, suitability may be demonstrated by accreditation under the Lighting Industry Commissioning Scheme.

4.47 Ductwork leakage testing should be carried out on new or refurbished ducting where practicable in accordance with the procedures set out in HVCA DW/143[16] on systems served by fans with a design flow rate greater than 1 m^3/s and for those sections of ductwork where the pressure class is such that DW/143 recommends testing.

Membership of the HVCA specialist ductwork group or the Association of Ductwork Contractors and Allied Services could be a way of demonstrating suitable qualifications for this testing work.

4.48 If a ductwork system fails to meet the leakage standard, remedial work should be carried out as necessary to achieve satisfactory performance in retests and further ductwork sections should be tested as set out in DW/143.

---

[16] DW/143 *A Practical Guide to Ductwork Leakage Testing*, HVCA, 2000.

Section 5: Guidance on thermal elements

THE PROVISION OF THERMAL ELEMENTS

5.1 New *thermal elements* must comply with paragraph L1(a)(i) of Schedule 1 to the Building Regulations. Work on existing *thermal elements* must comply with regulation 4A of the Building Regulations.

4A.–(1) Where a person intends to renovate a thermal element, such work shall be carried out as is necessary to ensure that the whole thermal element complies with the requirements of paragraph L1(a)(i) of Schedule 1.

(2) Where a thermal element is replaced, the new thermal element shall comply with the requirements of paragraph L1(a)(i) of Schedule 1.

U-values

5.2 U-values shall be calculated using the methods and conventions set out in BR 443.

5.3 Reasonable provision for newly constructed *thermal elements* such as those constructed as part of an extension would be to meet the standards set out in Table 4.

5.4 Reasonable provision for those *thermal elements* constructed as replacements for existing elements would be to meet the standards set out in Table 4.

*Curtain walling is treated as a **controlled fitting** and guidance is given in paragraph 4.28.*

Table 4 Standards for new *thermal elements*

Element[1]	Standard (W/m².K)
Wall	0.28[2]
Pitched roof – insulation at ceiling level	0.16
Pitched roof – insulation at rafter level	0.18
Flat roof or roof with integral insulation	0.18
Floors[3]	0.22[4]
Swimming pool basin	0.25[5]

Notes:

1. 'Roof' includes the roof parts of dormer windows, and 'wall' includes the wall parts (cheeks) of dormer windows.
2. A lesser provision may be appropriate where meeting such a standard would result in a reduction of more than 5% in the internal floor area of the room bounded by the wall.
3. The U-value of the floor of an extension can be calculated using the exposed perimeter and floor area of the whole enlarged building.
4. A lesser provision may be appropriate where meeting such a standard would create significant problems in relation to adjoining floor levels.
5. See paragraph 4.14.

Continuity of insulation and airtightness

5.5 The building fabric should be constructed so that there are no reasonably avoidable thermal bridges in the insulation layers caused by gaps within the various elements, at the joints between elements, and at the edges of elements such as those around window and door openings. Reasonable provision should also be made to reduce unwanted air leakage through the new envelope parts. The work should comply with all the requirements of Schedule 1, but particular attention should be paid to Parts F and J.

5.6 Significant reductions in thermal performance can occur where the air barrier and the insulation layer are not contiguous and the cavity between them is subject to air movement. To avoid this problem, either the insulation layer should be contiguous with the air barrier at all points in the building envelope, or the space between them should be filled with solid material such as in a masonry wall.

5.7 A suitable approach to showing the requirement has been achieved would be to submit a report signed by a suitably qualified person confirming that appropriate design details and building techniques have been specified, and that the work has been carried out in ways that can be expected to achieve reasonable conformity with the specifications. Reasonable provision would be to:

a. adopt design details published on the Accredited Construction Details website; or

b. demonstrate that the specified details provide adequate protection against surface condensation using the guidance in IP 1/06[17] and BR 497[18].

RENOVATION OF THERMAL ELEMENTS

5.8 For the purposes of this Approved Document, renovation of a thermal element through:

a. the provision of a new layer means either of the following activities:

   i. Cladding or rendering the external surface of the thermal element; or

   ii. Dry-lining the internal surface of a thermal element.

---

[17] IP 1/06 *Assessing the effects of thermal bridging at junctions and around openings in the external elements of buildings*, BRE 2006.
[18] BRE Report BR 497 *Conventions for calculating linear thermal transmittance and temperature factors*, 2007.

GUIDANCE ON THERMAL ELEMENTS

b. the replacement of an existing layer means either of the following activities:

   i. stripping down the element to expose the basic structural components (brick/blockwork, timber/metal frame, joists, rafters, etc.) and then rebuilding to achieve all the necessary performance requirements. As discussed in paragraph 3.9, particular considerations apply to renovating elements of traditional construction; or

   ii. replacing the water proof membrane on a flat roof.

5.9 Where a *thermal element* is subject to a *renovation* through undertaking an activity listed in paragraph 5.8a or 5.8b, the performance of the whole element should be improved to achieve or better the relevant U-value set out in column (b) of Table 5, provided the area to be renovated is greater than 50 per cent of the surface of the individual element or 25 per cent of the total building envelope. When assessing this area proportion, the area of the element should be taken as that of the individual element, not all the elements of that type in the building. The area of the element should also be interpreted in the context of whether the element is being renovated from inside or outside, e.g. if removing all the plaster finish from the inside of a solid brick wall, the area of the element is the area of external wall in the room. If removing external render, it is the area of the elevation in which that wall sits.

This means that if all the roofing on the flat roof of an extension is being stripped down, the area of the element is the roof area of the extension, not the total roof area of the dwelling. Similarly, if the rear wall of a single-storey extension was being re-rendered, it should be upgraded to the standards of Table 5 column (b), even if it was less than 50 per cent of the total area of the building elevation when viewed from the rear. If plaster is being removed from a bedroom wall, the relevant area is the area of the external wall in the room, not the area of the external elevation which contains that wall section. This is because the marginal cost of dry-lining with insulated plasterboard rather than plain plasterboard is small.

5.10 If achievement of the relevant U-value set out in column (b) of Table 5 is not technically or functionally feasible or would not achieve a *simple payback* of 15 years or less, the element should be upgraded to the best standard that is technically and functionally feasible and which can be achieved within a *simple payback* of no greater than 15 years. Guidance on this approach is given in Appendix A to Approved Document L1B.

5.11 When renovating thermal elements, the work should comply with all the requirements in Schedule 1, but particular attention should be paid to Parts F and J.

RETAINED THERMAL ELEMENTS

5.12 Part L of Schedule 1 to the Building Regulations applies to *thermal elements* in the following circumstances:

a. where an existing *thermal element* is part of a building subject to a material change of use;

b. where an existing element is to become part of the thermal envelope, where previously it was not;

c. where an existing element is being upgraded as a *consequential improvement* (regulation 17D) in accordance with paragraphs 6.1 to 6.11.

5.13 Reasonable provision would be to upgrade those *thermal elements* whose U-value is worse than the threshold value in column (a) of Table 5 to achieve the U-value given in column (b) of Table 5, provided this is technically, functionally and economically feasible. A reasonable test of economic feasibility is to achieve a *simple payback* of 15 years or less. Where the standard given in column (b) is not technically, functionally or economically feasible, then the element should be upgraded to the best standard that is technically and functionally feasible and which meets a *simple payback* criterion of 15 years or less. Generally, this lesser standard should not be worse than 0.7 W/m^2.K.

Examples of where lesser provision than column (b) might apply are where the thickness of the additional insulation might reduce usable floor area by more than 5 per cent or create difficulties with adjoining floor levels, or where the weight of the additional insulation might not be supported by the existing structural frame.

GUIDANCE ON THERMAL ELEMENTS

Table 5 Upgrading retained thermal elements

Element[1]	U-value W/m².K	
	(a) Threshold	(b) Improved
Wall – cavity insulation	0.70	0.55[2]
Wall – external or internal insulation	0.70	0.30[3]
Floors[4,5]	0.70	0.25
Pitched roof – insulation at ceiling level	0.35	0.16
Pitched roof – insulation at rafter level[6]	0.35	0.18
Flat roof or roof with integral insulation[7]	0.35	0.18

Notes:

1. 'Roof' includes the roof parts of dormer windows, and 'wall' includes the wall parts (cheeks) of dormer windows.
2. This applies only in the case of a cavity wall capable of accepting insulation. Where this is not the case it should be treated as for 'wall – external or internal insulation'.
3. A lesser provision may be appropriate where meeting such a standard would result in a reduction of more than 5% in the internal floor area of the room bounded by the wall.
4. The U-value of the floor of an extension can be calculated using the exposed perimeter and floor area of the whole enlarged building.
5. A lesser provision may be appropriate where meeting such a standard would create significant problems in relation to adjoining floor levels.
6. A lesser provision may be appropriate where meeting such a standard would create limitations on head room. In such cases, the depth of the insulation plus any required air gap should be at least to the depth of the rafters, and the thermal performance of the chosen insulant should be such as to achieve the best practicable U-value.
7. A lesser provision may be appropriate if there are particular problems associated with the load-bearing capacity of the frame or the upstand height.

5.14 When renovating *thermal elements*, the work should comply with all the requirements in Schedule 1, but particular attention should be paid to Parts F and J.

Section 6: Consequential improvements

6.1 Regulation 17D of the Building Regulations may require additional work to be undertaken to make an existing building more energy efficient when certain types of building work are proposed.

6.2 This requirement arises in existing buildings with a ***total useful floor area*** of over 1,000 m² where the proposed work consists of or includes:

a. an extension;

b. the initial provision of any ***fixed building service*** (other than a renewable energy generator);

c. an increase to the installed capacity of any ***fixed building service*** (other than a renewable energy generator).

6.3 Where regulation 17D applies, ***consequential improvements***, in addition to the proposed building work (the ***principal works***), should be made to ensure that the building complies with Part L, to the extent that such improvements are technically, functionally and economically feasible. Paragraphs 6.4 to 6.11 below set out guidance on what will constitute technically, functionally and economically feasible ***consequential improvements*** in various circumstances.

*The **principal works** must comply with the **energy efficiency requirements** in the normal way.*

6.4 Where improvement works other than the 'trigger activities' listed in regulation 17D (1) are planned as part of the building work, owners can use these as contributing to the ***consequential improvements***. The exception to this is if additional work is being done to the existing building to compensate for a poorer standard of an extension (see paragraphs 4.9 to 4.11).

*For example, if, as well as extending the building, the proposals included total window replacement, then the window replacement work would satisfy the requirement for **consequential improvements**, provided the cost was at least 10 per cent of the cost of the extension.*

6.5 Measures such as those listed in Table 6 that achieve a ***simple payback*** not exceeding 15 years will be economically feasible unless there are unusual circumstances.

For example, if the remaining life of the building is less than 15 years it would be economic to carry out only improvements with payback periods within that life.

Table 6 Improvements that in ordinary circumstances are practical and economically feasible

Items 1 to 7 will usually meet the economic feasibility criterion set out in paragraph 6.5. A shorter payback period is given in item 8 because such measures are likely to be more capital intensive or more risky than the others.

No.	Improvement measure
1	Upgrading heating systems more than 15 years old by the provision of new plant or improved controls
2	Upgrading cooling systems more than 15 years old by the provision of new plant or improved controls
3	Upgrading air-handling systems more than 15 years old by the provision of new plant or improved controls
4	Upgrading general lighting systems that have an average lamp efficacy of less than 40 lamp-lumens per circuit-watt and that serve areas greater than 100 m² by the provision of new luminaires or improved controls
5	Installing energy metering following the guidance given in CIBSE TM 39
6	Upgrading ***thermal elements*** which have U-values worse than those set out in column (a) of Table 5 following the guidance in paragraphs 5.12 and 5.13
7	Replacing existing windows, roof windows or rooflights (but excluding display windows) or doors (but excluding high-usage entrance doors) which have a U-value worse than 3.3 W/m².K following the guidance in paragraphs 4.23 to 4.28
8	Increasing the on-site low and zero carbon (LZC) energy-generating systems if the existing on-site systems provide less than 10% of on-site energy demand, provided the increase would achieve a simple payback of 7 years or less
9	Measures specified in the Recommendations Report produced in parallel with a valid Energy Performance Certificate

Consequential improvements on extending a building

Constructing a new free-standing building on an existing site (e.g. a new out-patients building at an existing hospital site, or a new classroom block at a school) is not an extension. These should be treated as new buildings.

6.6 Where a building is extended, or the habitable area is increased, a way of complying with regulation 17D would be to adopt measures such as those in Table 6 to the extent that their value is not less than 10 per cent of the value of the ***principal works***. The value of the ***principal works*** and the value of the ***consequential improvements*** should be established using prices current at the date the proposals are made known to the ***BCB***. They should be made known by way of a report signed by a suitably qualified person as part of the initial notice or deposit of plans.

An example of a suitably qualified person would be a chartered quantity surveyor.

CONSEQUENTIAL IMPROVEMENTS

Consequential improvements on installing building services

6.7 Where it is proposed to install a *fixed building service* as a first installation, or as an installation which increases the installed capacity per unit area of an existing service, reasonable provision would be to:

a. firstly improve the fabric of those parts of the building served by the service, where this is economically feasible; and

*This means for example that if heating systems are to be installed for the first time in a building or part thereof, or the installed heating capacity per unit area of an existing system is to be increased, the fabric should be improved. The aim in these cases is to make cost-effective improvements to the performance of the fabric so that the installed capacity (and the initial cost) of the **fixed building services** and their subsequent energy consumption are not excessive.*

b. additionally, make improvements in line with the guidance in paragraph 6.6. The cost of any improvement made as a result of following the guidance in sub-paragraph a) above cannot be taken as contributing to the value of the *consequential improvements* specified in paragraph 6.3.

If only the improvements under a) were made, then the CO_2 emissions from the building might well increase as a result of the higher level of servicing. By also requiring the general improvements in b), an overall improvement should be achieved.

6.8 For the purposes of this Approved Document, the installed capacity of a *fixed building service* is defined as the design output of the distribution system output devices (the terminal units) serving the space in question, divided by the *total useful floor area* of that space.

*This means that if (e.g.) the size of central boiler plant is increased to serve a new extension rather than to increase the heating provision in the existing building, the **consequential improvements** in paragraph 6.6 would be required but those in the following paragraphs would not apply.*

6.9 Reasonable provision for improving the fabric of those parts of the building served by the service in line with paragraph 6.7a above would be to follow the guidance in paragraphs 6.10 and 6.11 to the extent that the work is technically, functionally and economically feasible. The extent of such work is not limited by the 10 per cent threshold. The following paragraphs give guidance on what in normal circumstances would be economically feasible.

6.10 Where the installed capacity per unit area of a heating system is increased:

a. the *thermal elements* within the area served which have U-values worse than those set out in column (a) of Table 5 should be upgraded following the guidance in paragraphs 5.12 and 5.13; and

b. existing windows, roof windows or rooflights (but excluding *display windows*) or doors (but excluding *high-usage entrance doors*) within the area served and which have U-values worse than 3.3 W/m^2.K should be replaced following the guidance in paragraphs 4.23 to 4.28.

6.11 Where the installed capacity per unit area of a cooling system is increased:

a. *thermal elements* within heated areas which have U-values worse than those set out in column (a) of Table 5 should be upgraded following the guidance in paragraphs 5.12 and 5.13; and

b. if the area of windows, roof windows (but excluding *display windows*) within the area served exceeds 40 per cent of the façade area or the area of rooflights exceeds 20 per cent of the area of the roof and the design solar load exceeds 25 W/m^2, then the solar control provisions should be upgraded such that at least one of the following four criteria is met:

   i. the solar gain per unit floor area averaged over the period 0630 to 1630 GMT is not greater than 25 W/m^2 when the building is subject to solar irradiances for July as given in the table of design irradiancies in CIBSE Design Guide A;

   ii. the design solar load is reduced by at least 20 per cent;

   iii. the effective g-value is no worse than 0.3;

   iv. the zone or zones satisfies the criterion 3 check in Approved Document L2A based on calculations by an approved software tool; and

This will reduce the solar gain and hence the space cooling demand. Section 5.1 of TM 37[19] gives guidance on calculating solar gains, and Sections 4.4 and 4.5 give guidance on the effective g-value.

c. any general lighting system within the area served by the relevant *fixed building service* which has an average lamp efficacy of less than 45 lamp-lumens per circuit-watt should be upgraded with new luminaires and/or controls following the guidance in the *Non-Domestic Building Services Compliance Guide.*

This will reduce the lighting load and hence the space cooling demand.

---

[19] TM 37 *Design for improved solar shading control*, CIBSE, 2006.

Section 7: Providing information

7.1 On completion of the work, in accordance with paragraph L1(c) of Schedule 1, the owner of the building should be provided with sufficient information about the building, the *fixed building services* and their operating and maintenance requirements so that the building can be operated in such a manner as to use no more fuel and power than is reasonable in the circumstances. This requirement applies only to the work that has actually been carried out – e.g. if the work involves replacing windows, there is no obligation on the contractor to provide details on the operation of the heating system.

Building log book

7.2 A way of showing compliance with the requirement would be to produce the necessary information following the guidance in CIBSE TM 31 *Building log book toolkit*[20], or to add it to an existing log book. If an alternative guidance document is followed in preparing the log book, then the information conveyed and the format of presentation should be equivalent to TM 31.

7.3 The information should be presented in templates as or similar to those in TM 31. The information should be provided in summary form, suitable for day-to-day use. It could draw on or refer to information available as part of other documentation, such as the Operation and Maintenance Manuals and the Health and Safety file required by the CDM Regulations.

7.4 The new or updated log book should provide details of:

a. any newly provided, renovated or upgraded *thermal elements* or *controlled fittings*;

b. any newly provided *fixed building services*, their method of operation and maintenance;

c. any newly installed energy meters; and

d. any other details that collectively enable the energy consumption of the building and building services constituting the works to be monitored and controlled.

---

[20] TM 31 *Building log book toolkit*, CIBSE, 2006.

Appendix A: Documents referred to

BRE
www.bre.co.uk

BR 443 Conventions for U-value calculations, 2006. (Downloadable from www.bre.co.uk/uvalues)

Information Paper IP1/06 Assessing the effects of thermal bridging at junctions and around openings in the external elements of buildings, 2006. ISBN 978 1 86081 904 9

BRE Report BR 497 Conventions for Calculating Linear Thermal Transmittance and Temperature Factors, 2007. ISBN 978 1 86081 986 5

CIBSE
www.cibse.org

TM 31 Building Log Book Toolkit, CIBSE 2006. ISBN 978 1 90328 771 2

TM 37 Design for improved solar shading control, 2006. ISBN 978 1 90328 757 6

TM 39 Building energy metering, 2010. ISBN 978 1 90684 611 4

TM 46 Energy benchmarks, CIBSE 2008.

CIBSE Commissioning Code M: Commissioning management, CIBSE 2003. ISBN 978 1 10328 733 0

Department for Business, Innovation and Skills
www.bis.gov.uk

Technical Standards and Regulations Directive 98/34/EC (Available at www.bis.gov.uk/policies/innovation/infrastructure/standardisation/tech-standards-directive)

Department for Energy and Climate Change (DECC)
www.decc.gov.uk

The Government's Standard Assessment Procedure for energy rating of dwellings, SAP 2009. (Available at www.bre.co.uk/sap2009)

Current Energy Prices (www.decc.gov.uk/en/content/cms/statistics/publications/prices/prices.aspx)

English Heritage
www.english-heritage.org.uk

Building Regulations and Historic Buildings, 2002 (revised 2004) and other guidance

Health and Safety Executive (HSE)
www.hse.gov.uk

L24 Workplace Health, Safety and Welfare: Workplace (Health, Safety and Welfare) Regulations 1992, Approved Code of Practice and Guidance, The Health and Safety Commission, 1992. ISBN 978 0 71760 413 5

Heating and Ventilating Contractors Association
www.hvca.org.uk

DW/143 A practical guide to ductwork leakage testing, HVCA 2000. ISBN 978 0 90378 330 9

NBS (on behalf of Department for Communities and Local Government)
www.thebuildingregs.com

Non-Domestic Building Services Compliance Guide, CLG 2010. (Available to download from www.planningportal.gov.uk)

Legislation

SI 1991/1620 Construction Products Regulations 1991

SI 1994/3051 Construction Products (Amendment) Regulations 1994

SI 1994/3260 Electrical Equipment (Safety) Regulations 1994

SI 2000/2531 The Building (Approved Inspectors etc.) Regulations 2000

SI 2000/2532 The Building (Approved Inspectors etc.) Regulations 2000

SI 2006/3418 Electromagnetic Compatibility Regulations 2006

Decision No 1/95 of the EC-Turkey Association Council of 22 December 1995

Appendix B: Standards referred to

BS EN ISO 13370:2007 Thermal performance of buildings. Heat transfer via the ground. Calculation methods.

BS EN 14351-1:2006 Windows and doors. Product standard, performance characteristics. Windows and external pedestrian doorsets without resistance to fire and/or smoke leakage characteristics.

Index

A

Accredited construction details schemes 5.7
Agricultural buildings 3.15–3.16
Air leakage testing (ductwork) 4.47–4.48
Airtightness 5.5–5.7
Ancient monuments 3.5, 3.8
Approved Document L2B
 Conventions 1.9
 Purpose 1.1–1.5
 Types of work covered 3.2–3.4
Architectural interest 3.8
Area-weighted U-value 4.7–4.8

B

BCB
 See Building Control Body (BCB)
British Standards Appendix B
 BS EN 14351-11 4.25
 BS EN ISO 13370 4.14
Building Control Body (BCB)
 Definition 3.1
 Notice of completion of commissioning 4.41–4.48
 Notification of work 3.23–3.28
Building fabric
 U-values 4.3
 See also Thermal elements
Building log book 7.2–7.4
Building services
 See Fixed building services
Business park units 3.4

C

Cavity walls Table 5
CE marking 3.32
Certification
 Competent person self-certification schemes 3.25–3.28
 Product certification schemes 3.34–3.36
Change of energy status 3.20, 4.16–4.21
Change of use 4.15
Cladding 5.8
Commissioning
 Definition 3.1
 Fixed building services 3.30, 4.36–4.48
Community energy systems 4.32
Competent person self-certification schemes 3.25–3.28
Compliance with requirements
 Materials and workmanship 3.27, 3.36
 Self-certification 3.28
Condensation risk 4.9
Consequential improvements 3.4, 3.20, 6.1–6.11, Table 6
 Definition 3.1
 Extensions 4.1, 4.10
Conservation areas 3.5
Conservatories 3.21–3.22, 4.12–4.13
Controlled fittings 4.19, 4.23–4.28
 Definition 3.1
 See also Doors; Windows
Controlled services 4.19, 4.29–4.35
 Definition 3.1
 See also Fixed building services
Control systems
 Fixed building services 4.30
Curtain walling 4.28

D

Design flexibility 4.6–4.11, 4.21

Display windows 4.19
 Definition 3.1
Doors
 Area of opening 4.19, Table 2
 U-values Table 3, 4.19, 4.23–4.25, 4.27
Dry lining 5.8
Ductwork
 Air leakage testing 4.47–4.48
Dwellings
 Definition 3.1
 Mixed-use buildings 3.4

E

Emergency escape lighting
 Definition 3.1
Emergency repairs 3.24
Energy efficiency requirements 2.1
 Definition 3.1
 Exemptions 3.5
 Fixed building services 4.30
Energy meters 4.33–4.35
Energy Performance Certificate 2.1
Energy status change 3.20, 4.16–4.21
European standards 3.32, 3.40
Exemptions 3.5
Extensions 3.4, 4.1–4.14
 Consequential improvements 6.6
 Historic or traditional buildings 3.11

F

Fabric
 See Building fabric
Fit-out works 3.4
 Definition 3.1
Fixed building services
 Commissioning 3.30, 4.36–4.48
 Consequential improvements 6.7–6.11
 Definition 3.1
 Extensions 4.5
 System efficiencies 4.30
 See also Controlled services; Heating and hot water systems; Lighting
Flat roofs 5.8, 5.9, Table 5
Floors Table 4, Table 5

G

Glazing
 See Windows

H

Health and safety 3.41
Heating and hot water systems 4.29
 Consequential improvements 6.7–6.8
 Controls 4.30
 Non-notifiable work 3.30
 See also Fixed building services
High-usage entrance doors 4.19, Table 3
 Definition 3.1
Historic buildings 3.7–3.13

I

Industrial buildings
 Exemptions 3.15–3.16
Information provision 7.1–7.4
Institutional (residential) buildings 3.2
Insulation
 See Thermal insulation
Internal lighting 4.29, 6.11

L

Legislation Appendix A
 European 3.32
Lighting 4.29, 6.11
Limitation on requirements 2.2–2.3
Listed buildings 3.5
Low energy demand 3.6, 3.15–3.20

M

Maintenance instructions 7.1
Material change of use 4.15
Materials and workmanship 3.31–3.33
Minor works 3.24
Modular buildings 3.4, 3.6

N

National parks 3.8
Non-exempt buildings 3.17–3.20
Non-notifiable work 3.29–3.30
Notice of completion of commissioning 4.41–4.48
Notification of work 3.23–3.30

O

Operating and maintenance instructions 7.1

P

Party walls 4.19
Passive control measures 6.11
Payback period 5.10, 5.13, 6.5
Places of worship 3.14
Porches 3.21–3.22, 4.12–4.13
Portable buildings 3.4, 3.6
Principal works
 Definition 3.1
Product certification schemes 3.34–3.36
Publications (excluding BSI and European Standards) Appendix A
 Assessing the effects of thermal bridging at junctions and around openings in the external elements of buildings (IP 1/06, 2006) 5.7
 Building energy metering (CIBSE TM39, 2010) 4.34
 Building log book toolkit (CIBSE TM31, 2006) 7.2
 Commissioning management (CIBSE Code M, 2002) 4.42
 Conventions for calculating linear thermal transmittance and temperature factors (BRE 497, 2007) 5.7
 Energy benchmarks (CIBSE TM46, 2008) 4.35
 Non-Domestic Building Services Compliance Guide (CLG, 2010) 3.18, 4.29
 Practical Guide to Ductwork Leakage Testing (HVCA DW/143, 2000) 4.47

R

Rendering 5.9
Renewable energy systems 4.31
Renovation
 Definition 3.1
 Thermal elements 5.8–5.11
Replacement
 Appliances 3.30, 4.30
 Thermal elements 5.4, 5.8
 Windows 4.24, 6.4

L2B INDEX

Replastering 5.9
Rooflights 4.4, 4.23–4.27, Table 2, Table 3
Roofs Table 4, Table 5
Roof ventilators Table 3
Roof windows 4.23–4.28, Table 3
Rooms for residential purposes 3.2

S

Self-certification schemes 3.25–3.28
Shell and core developments 3.4
Simple payback 5.10, 5.13, 6.5
    Definition 3.1
Solar control 6.11
Standards 3.37–3.40
    European 3.32, 3.40
Swimming pools 4.14, Table 4

T

Technical specifications 3.37–3.40
Temporary buildings 3.5, 3.6
Thermal bridges 5.5
Thermal elements 5.1–5.14
    Consequential improvements 6.9–6.11
    Definition 3.1
    Renovation 5.8–5.11
    Replacement 5.4, 5.8
    Retained 4.19, 5.12–5.13
    Upgrades 4.19, 5.13, 6.10, 6.11, Table 5
    U-values 5.2–5.4
Thermal insulation 3.30, 5.5–5.7
Thermal separation 4.12
Total useful floor area 3.5, 4.1–4.2
    Definition 3.1
Traditional construction 3.7–3.13

U

Unheated buildings 3.15
Upgrades 4.11
    Lighting 6.11
    Thermal elements 4.19, 5.13, 6.10, 6.11, Table 5
U-values
    Area-weighted 4.7–4.8
    Controlled fittings 4.25–4.28, Table 3
    Thermal elements 5.2–5.4, Table 4, Table 5

W

Walls Table 4, Table 5
Whole building calculation method 4.9–4.11
Windows 4.23–4.28
    Area of opening 4.4, 4.19, Table 2
    Replacement 4.24, 6.4
    U-values 4.19, 4.25–4.27, Table 3
Workmanship 3.31–3.33
Workplace (Health, Safety and Welfare) Regulations 1992 3.41
Workshops 3.15–3.16

Z

Zoned controls 4.30